John Simpkins Butler

The Curability of Insanity and the Individualized Treatment

of the Insane

John Simpkins Butler

The Curability of Insanity and the Individualized Treatment of the Insane

ISBN/EAN: 9783337372903

Printed in Europe, USA, Canada, Australia, Japan

Cover: Foto ©berggeist007 / pixelio.de

More available books at **www.hansebooks.com**

THE

CURABILITY OF INSANITY

AND THE

INDIVIDUALIZED TREATMENT OF THE INSANE

BY

JOHN S. BUTLER, M.D.

HARTFORD, CONN.

LATE PHYSICIAN AND SUPERINTENDENT OF THE CONNECTICUT RETREAT FOR THE
INSANE ; MEMBER OF THE CONNECTICUT STATE BOARD OF HEALTH ;
HONORARY MEMBER OF THE MEDICO-PSYCHOLOGICAL
ASSOCIATION OF GREAT BRITAIN

NEW YORK AND LONDON

G. P. PUTNAM'S SONS

The Knickerbocker Press

1887

INSANITY is a disease of the brain, including a departure from ordinary modes of thought and states of feeling in health.—*Dr. Ray.*

INSANITY is a calamity incident alike to tender sensibility, to grand enthusiasm, to sublime genius, and to intense exertion of the intellect.—*Sir James Macintosh.*

WHOEVER has brought himself to consider a disease of the brain as differing only in a degree from a disease of the lung, has robbed insanity of that mysterious horror which forms its chief malignity. —*Sir James Macintosh—Life of Robert Hall.*

THE physician, confident in the assurance that patient and careful observation of insanity, with the earnest desire to understand its nature, does fit him to express with authority the results of his experience, must not shrink from pronouncing his opinion sincerely and fearlessly, however unpopular it may be.—*Maudesley on Responsibility in Mental Disease.*

iii

THE CURABILITY OF INSANITY AND THE INDIVIDUALIZED TREATMENT OF THE INSANE.

I N the comparisons of the provisons made for the insane in the United States, in 1844, with those of the
present day, we find the best measure of progress to be in
the larger recognition of their necessities, in remedial
treatment, custodial provision, and acceptance of the
power of prevention as applicable to insanity as to other
physical diseases.

In October, 1844, thirteen gentlemen met in Philadelphia and organized the " Association of Medical Superintendents of Institutions for the Insane." Their object
was, by a comparison of views and careful study, " to
secure for the future a higher standard for hospitals,
and a more liberal and enlightened treatment for all suffering from mental diseases." The causes which led to
this result are stated in the Secretary's history of the
Association. At that time, 1844, there were in the United
States twenty-five lunatic hospitals of all classes, containing less than twenty-six hundred or twenty-seven hundred
inmates. The largest number in a distinct hospital was
two hundred and sixty-three, in that of Worcester, although there were three hundred and fifty in the Recep-

tacle on Blackwell's Island.   According to the report of
the Board of the State Charities in Pennsylvania, in September, 1883, there were in the United States one hundred and forty-seven lunatic asylums, containing fifty-
one thousand eight hundred and seventeen patients ; the
total number of insane in the United Sates being esti-
mated to be ninety-two thousand, or one in five hundred
and forty-five of the population, the lowest rate of insanity
being found in the more recently settled States.   The
Association at this time embraces all North American
institutions, and now records one hundred and twenty-
two active and retired members.   Well may the excellent
and most efficient Secretary say of the Association that,
" Formed in the interest and for the promotion of the
welfare of the insane, it has been steadily growing in
numbers, in influence and power, until it covers with
its protecting shield a large proportion of the insane
throughout the length and breadth of the land."   In the
eventful history of the Association for the past forty years
there has been, for the most part, a singular and cordial
unanimity of action as to the best means of attaining the
desired end—the highest good of the insane. The " Propositions " adopted by the Association show not only
a large wisdom, but a foresight of the necessities of this
comparatively new and unexplored field of philanthropy.
   The unexpectedly large and continually increasing
number demanding either hospital treatment, or simply
hospital supervision and care, has naturally led to a diversity of opinion as to the number of patients that can be
most profitably treated in one institution. That the causes
of this diversity may be better understood, and my own

position more clearly defined, I may here quote some of these propositions, and my reasons for objecting, not only to the one accepted by a close vote, but to the others subsequently passed in accordance with it.

At the meeting in Philadelphia, in 1851, among other propositions, the following was unanimously adopted:

The highest number that can, with propriety, be treated in one building is 250, while 200 is a preferable maximum.[1]

At the meeting in Washington, in 1866, the following propositions were adopted:

Insane persons considered incurable, and those supposed curable, should not be provided for in separate establishments.

*　　*　　*　　*　　*　　*

The large States should be divided into geographical districts of such size that a hospital, situated at or near the centre of the district, shall be practically accessible to all people living within its boundaries, . . . and available for their benefit in cases of mental disorder.

All State, county, and city hospitals for the insane should receive all persons belonging to the vicinage designed to be accommodated by such hospitals, who are afflicted with insanity, whatever may be the form of the bodily disease accompanying the mental disorder.

The enlargement of any such specified institution may be properly carried, as required, to the extent of accommodating six hundred patients, embracing the usual proportion of curable and incurable in a particular community.

---

[1] The *International Record* for April, 1887, gives the number of inmates in each of eighty-eight of our lunatic hospitals. Of these eighty-eight hospitals only sixteen contain not more than 250 patients, —that "highestnumber" accepted by the Association in 1851 ; thirty-eight contain from 250 to 600 ; twenty-four from 600 to 1,000 ; and ten from 1,000 to 1,818.

The complaint of the continued over-crowding for admission seems unabated.

While the other propositions were quite unanimously accepted, this was passed, near the close of the meeting, by a vote of eight to six.   Under the increasing pressure of necessary admissions, this proposition has seemed practically to annul any official limitation of number.

At the meeting at Toronto, in 1871, the Association reaffirmed, in the most emphatic manner, all the former declarations in regard to hospital organization, management, etc., and also,

RESOLVED, That neither humanity, economy, nor expediency can make it desirable that the care of the recent and chronic insane should be in separate institutions.

My own position upon this point of numbers was early taken, and I have seen no good reason to change.   At the meeting in Pittsburg, in 1865, I stated to the Association that the admission into the Retreat of a large number of incurable State patients had greatly embarrassed the remedial treatment of the recent and hopefully curable.   And I was, consequently, led to suggest the consideration by the meeting of some kind of distinct and efficient provision to be adopted by the State for these unfortunates.   I did this simply, without any distinctly formed plan of my own, but only to find the best effectual way of escaping from such possibly avoidable interference with hopefully curable treatment.   Most unexpectedly to me, this proposition led, as was reported, to the " most excitable debate of the session," as unexpectedly, to its almost unanimous disapproval, only one member (Dr. Hills, of Ohio) voting with me in favor of it.   I then offered the following motion :

RESOLVED, That a committee of three be appointed to take into consideration the condition of the chronic and supposed incurable insane, and the possible arrangement for their treatment and custody, and to report at the next meeting of the Association.

The motion was, in due courtesy, passed, and Drs. Butler, Walker, and Curwen were appointed the committee.

A long vacation, made necessary by illness, prevented both my preparation of a report and my attendance at the next meeting.

At the meeting in 1866, propositions favorable to my views were presented by Dr. Walker, and rejected ; while others of an opposite import, by Dr. Chipley, were accepted. The unanimous reaffirmation of all the propositions heretofore adopted, clearly defined the decision of the Association on those points.

I hold that these later propositions fail to anticipate the large increase of the number of the insane, the larger hospital accommodations they demand, and, especially, the changes so rapidly coming over the different classes. Neither the original thirteen, in 1844, nor the members who in 1851 voted that, " two hundred were a preferable maximum of inmates to be treated in one building," could have imagined the present number of insane, in and out of hospitals, or that its rapid increase would, in a single year (1884) add more than two hundred to their number in the single State of Massachusetts.

My proposition at the Pittsburg meeting in 1865 seemed to fall lifeless from the animated discussion which it had excited ; but the radical principle it contained, like good seed sown by more than one hand and in good ground, has, after twenty years of gradual and persistent development, come forward with better promise of acceptance in the future.

Here, fairly to myself, I may recall some of those events in my earlier professional life, which led me to determined opinions in regard to the necessities of the insane.

Early in 1833, shortly after I had commenced the practice of medicine in Worcester, Mass., I made a call simply of professional courtesy on Dr. Woodward, who had been lately appointed superintendent of the newly erected State Lunatic Hospital. While standing with him in the entrance hall, a party of his patients, "crazy men" (then a sadly strange sight to me), passed in from a walk. The Doctor stopped them to give an order to their attendant, and my attention was especially drawn to the pitiable appearance of the laggard of the group. Feeble and emaciated, he seemed to be a hopeless remainder of a man. The Doctor told me he was a young Welshman, Llewellyn by name, as I well remember, who had come to this country " to pick up gold in our streets." Unable to find work or wages, hearing sad news from his home in Wales, through homesickness he had sunk into the deepest melancholy. " Poor fellow," I said, " his is an utterly hopeless case." " By no means," answered Dr. Woodward. " But I mean *him*," pointing to Llewellyn, " he cannot recover ! " " I confidently expect he will," replied Dr. W. " May I see your treatment ? " I asked. " Every day, if you wish," was the Doctor's reply. For weeks following I saw him, if not every day, very frequently. On my return home I said to a friend :

"In my course of lectures in the Harvard Medical School, in my graduate and post-graduate courses in Philadelphia, I heard no such case described. In a library fairly well stocked for that day and faithfully consulted, no such case and treatment were given. If Llewellyn can be cured it will be next to a revelation in medicine to me." In a few weeks he came down to my office to bid me a grateful good-bye, etc., there presenting himself, in contrast with my first interview, a rarely good specimen of a healthy, vigorous, and intelligent young man. This case shaped the future of my professional life. For years afterwards I was a frequent visitor to the hospital and a somewhat careful observer in the wards, to all of which Dr. Woodward gave me free access. In those wards I saw frequent illustrations of the marvellous results of the *moral* treatment of the insane—that individualized power, which the healthy, intelligent, enthusiastic mind holds over the "untuned and jarring senses" of the lunatic. Then a young practitioner, striving to win public confidence and position, I found that I gave to my cases of typhus fever, etc., no more frequent, sharp, and kindly treatment than Dr. Woodward gave to his cases of recent insanity. This, especially, was before the first enlargement of the Hospital by the addition of two new wings, which Dr. Woodward greatly regretted, being confident it would cripple his system of treatment. He earnestly advised that it should be as an "annex" erected on the adjacent farm land of the Hospital.

In 1839 I was elected Resident Medical Officer of the Penal, Charitable, and Reformatory Institutions, and superintendent of the newly erected Lunatic Hospital, of the city of Boston. Those three years' superintendency of the Lunatic Hospital gave me the desired opportunity of applying to my own cases of insanity those principles of treatment which I had seen applied with such eminent success in Dr. Woodward's wards, a success which I have never seen surpassed, if equalled ; a fascinating illustration to me, of the merciful advances in these later days from the ignorance and cruel barbarism in the "mad houses" of "ye olden times," when, in the language of an old Scotch writer : "The better sort of ye mad people were given to the care of the chirurgeon, the baser sort, to the taming of the scourge !" See Appendix (1).

The *North American Review*, for January, 1843, has an article on "Insanity in Massachusetts." The writer says : "We select for description the Boston Lunatic Hospital in 1842. Its patients are wholly of the pauper class. Its inmates are of the worst and most hopeless class of cases. They are the raving madman and the gibbering idiot, whom, in the language of the inspectors of prisons, hospitals, etc., for Suffolk County, we had formerly seen tearing their clothes amid cold, lacerating their bodies, contracting most filthy habits, without self-control, unable to restrain the worst feelings, endeavoring to injure those who approached them, giving vent to their irritation in the most passionate, profane, and filthy language, fearing and feared, hating and almost hated. Now they are all neatly clad by day and comfortably lodged in separate rooms by night. They walk quietly with self-respect along their spacious and airy halls, or sit in listening groups around the daily paper, or they dig in the garden, or handle

edged tools, or stroll around the neighborhood with kind
and careful attendants. They attend daily and reverently upon religious exercises and make glad music with
their united voices. Such is the condition of the insane
of the city of Boston ; and although but twenty-eight out
of one hundred and seventy-one have been cured, and
the rest will probably wear out their lives in hopeless insanity, yet there is a melancholy pleasure in witnessing
the great amount of animal happiness they enjoy, in seeing the kind regard paid to prostrate humanity, the respect shown to the deserted temple of reason. It is only
as it were twining fresh flowers on the graves of the dead ;
still it is a grateful sight to the humane, and a more certain indication of high civilization, than the most refined
taste in literature and the arts, or the most fastidious of
social etiquette."

One of these patients came into the hospital out of an
iron cage, which I was told she had inhabited more than
a year, and several others out of veritable Barnum's me-
nagerie wooden cages. All these were females. Freed
from restraint and seclusion, soon after admission, they
were all readily won over to decent and orderly lives.
Before long all of them were occasional visitors in our
own family parlor.[1] I trust these details will not be con-

---

[1] Among the pauper lunatics admitted late in the autumn of 1839
into the recently organized Boston Lunatic Hospital, and whose condition was so graphically and truthfully described in the *North
American Review* for January, 1843, in the article on "Insanity in
Massachusetts," was a Scotch-Irish girl named Mary. She was a
fair specimen of the repulsive and difficult class which first occupied
those wards.

Mary was unusually athletic, naturally good-natured, but trained to
the necessity of fighting her own way, and yielding only to brute

sidered outside of good taste, as they seem to me to be but fairly descriptive of the natural outgrowth of due sympathy with the insane, as instructed and fortified by the teachings and examples of Dr. Woodward and others.

On my election, early in 1843, to the superintendency of the Connecticut Retreat for the Insane, I found that the directors had most charitably voted to admit, and at very low rates, many of the pauper and chronic insane from the almshouses of the State. In these, the earlier days of my Retreat life, when our crowded wards crip-

---

force. Her untamed and belligerent state was the natural result of her sadly neglected mental condition.

One day, passing by the outer door of her ward, my attention was attracted by the unusual brightness of a dandelion flower which on that cold autumn day had crowded its unseasonable (but as it proved its timely) way up through the lately graded ground. As I picked it up I was startled by a wild uproar within the ward. Hastening in, I met the frightened and then utterly inexperienced nurse, and, apprehending some serious trouble, sent her to hurry up an assistant officer to my possibly needed aid. I have never imagined a more striking illustration of an old time mad-house than that crowded hall then presented to me. Mary, her stout arms akimbo, was marching down the hall, her voice at the highest pitch, profanely denouncing everybody and every thing ; the other patients highly excited, some running for shelter, others joining the wild uproar. I announced my advent by a decided stamp on the floor, with an equally decided voice, saying : " I will not have such a tumult here." " The h—l you won't ! we 'll see to that," I very distinctly remember was her prompt answer, as, suddenly wheeling, she advanced to the decision of that question by " force of arms,"—arms to which mine would be but feeble impediments ! How heartily did I then wish myself back in general practice, out among the sick and wounded Irish *men* on the railroads ! But seemingly the supremacy of the week-old superintendent was then and there to be fought out on the instant ! Recalling the experience of my boyhood when, wrestling with the stouter farmer boys, I learned that a sudden catch at the collar and as quick a rap at the heels gave me the right relative position in the downfall, I waited, silently and quietly watching her ferociously maddened approach ! When just within reach suddenly her whole aspect changed

pled my means of classification, a quiet and apparently inoffensive case of dementia was necessarily located in one of the better wards ; the poor man would sit silent all day in a dreamy, stupid state, his only token of active life, the constant twirling of his thumbs. A refined and intelligent gentleman on the same hall, who was recovering from the results of an overworked brain, came to me one day, exclaiming with no little agitation, " Doctor, I *must* go home ! " I remonstrated, urging his rarely good prospects of a speedy recovery. " Why should you go ? "

---

and wonderfully softened. Naturally following her eye, and looking downward, for the ally who had so suddenly come to my relief, I found my hand, unconsciously upraised for the encounter, held that dandelion ! I offered it to her in a few low-toned words, and she accepted it with a very humble courtesy. The sudden and wonderful change of aspect was only surpassed by the touching pathos with which the subdued maniac exclaimed : " Oh ! beautiful ! can *you* mane to give *that* to the loikes of me ? " The battle was fought and won. Taking my offered arm, she went quietly to her room.

There was another chapter in the history of poor Mary, which may well be recorded. A few days after my instructive experience with the dandelion, she tore up the new dress given her on admission. After due reproof and persuasion, her promise of future good behavior was accepted, and another dress ordered. " Why, sir, she will certainly tear it up ! " said the nurse. " Then give her another, and, if need be, another," was my answer. Much to my disappointment, one or two more soon followed the first. My repeated expressions of surprise and sorrow, with kindly remonstrance against such unprovoked bad behavior, followed by seclusion in her room, and withdrawal of my daily and kindly personal attentions, soon produced the desired effect, namely, more self-control and a better mind. I then gave her a pretty dress pattern,—" good enough for a lady," she said,—and allowed her to help make it up. We all gave her a somewhat formal reception on her " coming out " in it ; and our comments were cordially, and with no little self-complacency, received. From that time she became quiet and industrious, an orderly attendant on chapel services, and at our little social parties.

Hopelessly insane, she remained in an orderly, submissive, and seemingly happy condition while under my protective care.

I asked. " Because," said he, " this continued rainy
weather has kept me in-doors for a fortnight. I am in
your way, in your business rooms. I have worn out the
hospitality of Mrs. Butler—up there, seeing that poor old
fellow ·twirling his thumbs hour after hour ! day after
day ! I can't stand it ! D—n it, I shall be just like
him ! " My continued experience ever afterwards
strengthened my convictions of the expediency and indeed humanity of the segregation of the chronic insane
from the recent and hopefully curable cases. I was
compelled by this conviction to present this question to
the consideration of the Association at the meeting at
Pittsburg, in 1865.

Individualized treatment is called for in insanity as
imperatively as in the case of acute forms of other physical disease. The form of treatment is different according
as the practitioner is hopefully working for a cure in an
acute case, or as in some chronic case of long standing,
he is simply administering palliation and general care.
The first requires his personal and persistent attention,
the second may be treated in a general way and may be
committed to others.

I believe strictly recent insanity in very many cases, is
radically curable under the prompt, persistent, and united
use of medical and moral means. These, to be efficient,
demand individualized application, *i. e.*, that same immediate, close, and sharp personal service which the general practitioner necessarily gives to the early stages of
typhus, diphtheria, cholera, etc. This power, essential to
the largest success, is limited, as in all individual efforts,
by number. Applicable to the few, it cannot be extended

to the many. While here and there it may reach one in a crowd, the general result proves the limitation. In addition to professional skill the largest success of individualism demands that combination of those innate, inherent qualities of "courage, kindness, and patience" which the *Edinburgh Review*, in April, 1814, attributed to William Tuke as the secret of his success in the Retreat for the Insane in York, a "success" which remains to this day our highest instruction ! As no two cases of insanity or physical disease are, in all causes and effects, precisely alike, the peculiarities of each will of course demand special consideration.

Dr. Conolly, in his admirable essay on the "Indications of Insanity," speaking of the duties of medical men, says (page 428) :

> To superintend with care and without offending ; to control without severity, and to indulge without weakness ; to attract without fatiguing the attention ; to revive the memory without reviving memorials of affliction ; to touch the imagination but not too sensibly ; to encourage at favorable moments to such comparisons as may triumph over retreating delusions, is a task too delicate, too sacred, I might say, to be entrusted to common hands. It is a power which *cannot be delegated.*[1]

---

[1] Esquirol relates that an old nun, formerly employed in the tuition of the young, was brought to the hospital in a state of profound melancholy. During six months all the methods which we had recourse to were without effect. Her notions always remained the same, and she continually repeated to the superintendent that he was wrong not to treat her as the most guilty of women, and that he ought to impose on her the severest punishments.

One day she met him in the interior of the hospital, and renewed the same kind of conversation. She received from him a sharp reply, with an express declaration that he would listen to her no more, since she always persisted in the same notions, and showed him no sort of confidence. The patient retired in silence to her apartment, re-

The best definition of the alleged power of individualism to charm down insanity, is that given by Emerson as
"The power *behind* the eye."

---

flected deeply on the reproof she had experienced, and did justice in
her own mind to the upright and irreproachable character of the superintendent, and his sincere desire to assist in the re-establishment of the
unhappy persons about him.   Was not all he had said dictated by
the most humane intentions?   She experienced great perplexity that
night, and a sort of interior conflict between the thought of her
imaginary crimes and the friendly remonstrance of a man who could
have no personal interest in what he said.   These vacillations, and
this interior conflict, going on for some time in the form of cool discussion in her own mind, produced the most favorable change in her,
and she ended by being fully convinced that her scruples were chi-
merical, and she at once entered eagerly upon the proper means for
the perfect re-establishment of her health by the proper physical
treatment.

The preceding case reminds me of one of my old patients whose
history on admission told of long time duration of persistent antagonism in the varied relations of her life.   She was a widow of middle
age, without living children ; of superior natural intellectuality and
refinement, and a quick pleasant manner.   Neglect in childhood, and
defective training in later life led to the natural consequences of the
undue indulgence of a strong will.   These, added to unhappy domestic
relations and *ill health*, readily induced insanity, though not of a
marked salient character.   In due time and manner I kindly and
frankly discussed with her the whole sad story of these unhappy
causes and effects, explaining their direct action upon her nervous
system.   I urged my hearty desire for her cordial acceptance of my
professional opinions, and her coöperation in what my every-day experience taught me was her only way of escape.   The result of thus
winning her confidence and sympathy, was a gradual but decided
gain in general health and better self-control.   Her "fallings from
grace" became less frequent and marked.   Some time afterward to
my great surprise and annoyance she came to me one day during a
temporary disturbance of her health, abruptly demanding an entire
change in my medical prescriptions.   I declined, pleasantly explaining the good working of the present remedy, and the peril of the extreme one she proposed.   She persisted in demanding the change.   I
sought to avoid the discussion, but in vain ; the difference of opinion
finally culminating in her saying with marked emphasis : " I know
my own case, sir, and I will have it "; and in my answering with all the
then possible courtesy : " Madam, you compel me to say you shall

The chapel,[1] the amusement hall, the social circle inside and the social circle *outside* of hospital wards have, in my experience, proved potent remedial agencies. Un-

---

not have it." She left the room with a defiant expression, which, shadowing my good hopes of her improvement, made me much solicitude. I found next morning a letter from her on my desk. I opened it with no little anxiety and reluctance, knowing the peremptory power of her pen. I expected a sharp continuance of her argument; but I gladly accepted her conclusion when I read,

"My dear, dear—Emperor Nicholas :—Forgive me—forgive me —I will try never to be so naughty again."

A young lady between sixteen and seventeen years of age, of good physical development, but apparently of a decidedly nervous temperament, lady-like in her present manners, was some time ago placed under my care. Her parents told me that she had been in pretty good health until within a year or two, when without any apparent cause she had become dyspeptic, oftentimes sleepless, generally unwell, and gradually growing worse and more nervous and excitable. She had of late become so uncontrollable that they were compelled to bring her to the Retreat, fearing she was drifting into insanity. As she was their only child, they now feared that they had been unwisely indulgent to her.

Her heartless indifference to the separation from her parents plainly indicated the "spoiled child."

A careful investigation of her case enabled me readily to detect the disordered condition of various organs, a condition easily preventible in its primary stages, and now as easily remedied by suita-

---

[1] "Though it is the first time for years that many of our inmates had been thus recognized as members of the human family, their fixed attention and serious deportment is a pleasant illustration of the adaptation of the Gospel to 'all sorts and conditions of men.' The utility of its influences should be undoubted. No one can look upon our household assembled for the instruction of the Sabbath, or for the family worship of the evening, and see them there as one family rise up silently and reverently to pray to 'Our Father in Heaven,' without realizing that some feel the solemnity of the act, without being convinced that a chord may be struck, whose ultimate vibration may awaken some recollection of early life and bring back upon the excited and bewildered mind some calm and solemn influences, and give that one moment of self-control in which the first link in the chain of diseased associations may be broken."—*Boston Lunatic Hospital Report, 1840.*

der such influences I have frequently detected the first indications of recovery. It seems self-evident to me that these details of moral treatment can be most successfully applied only in a hospital of the originally assumed "best number" of two hundred inmates. Here can be most easily developed those social influences which have such special power over diseases of the brain and nervous system. In such comparatively secluded lives, the natural cravings for sympathy and companionship most readily attract those genial affinities which lead to the formation of little homelike circles of newly found friends. The happiest results can often be traced from such circles. The reaction of mind upon mind, comparison and discussion, with criticisms, sometimes happily sharp they may be, but ever kindly, have here their

---

ble remedial agencies,—another of the familiar and sad old stories of neglected cause and effect.

As she appeared pleasant and lady-like, and quietly accepted her novel position, I located her with others of my lady patients, naturally expecting that the regaining of self-control would be aided by the influence of such refined and genial surroundings.

For a few days all went so well with her that I was beginning to think her parents had been needlessly alarmed. But the little stock of self-control was soon exhausted, even under the soothing influence of the new surroundings, when I was summoned hastily, the nurse reporting a violent outbreak of temper on slight provocation, and the free use of language "too bad to be repeated to you." On reaching the hall, assuming the appearance of great regret and mortification, I told the ladies that I came up to apologize for the sad blunder I had made in associating that "unfortunate young woman" with *them.* From the social position of her parents I had unfortunately taken it for granted that their daughter was a young lady, asking them to excuse my mistake, and directing "the young woman" to follow me. I immediately took her down to a lower department, occupied by a sadly repulsive class, whom the charity of our Board, before the erection of the State hospital, had temporarily received from the almshouses of the State. Bidding the nurse to be very patient with

place. The desponding are comforted and made hopeful, and the excited are repressed and instructed by the cordial and kindly comments of the convalescent and experienced. If all these desired good effects do not *immediately* result, they certainly remove at once that greatest evil of lunatic hospitals—monotony. Dr. Tuke, in his admirable "Illustrations of the Influence of the Mind upon the Body," speaking of the importance of this power as a practical remedy in disease, quotes John Hunter, as follows:

There is not a natural action in the body, whether voluntary or involuntary, that may not be influenced by the special state of the mind at the time.

The variety of ways by which one may promote the interest of the insane, happily illustrates the many-sidedness of truth.

---

this sad addition to her burdensome family, to shut her up in a room if her language became intolerable, I left her with no little anxiety as to the result of this "moral treatment."

In a brief time I was earnestly recalled, the nurse reporting that the young lady, in a paroxysm of tears, repentant of her folly, admitting my just decision, was begging the privilege of my forgiveness. Accepting her sincere repentance with great delight, and bidding the nurse to readjust her dress, and to remedy the flushed and tearful face with a little cold water, I gave the girl my arm and returned to the hall.. Entering the room with much assumed formality, I introduced my companion as "a charming young lady friend of mine, who had just come up to us from the country." The ladies readily appreciating our position, gave their returned associate a right merry and hearty reception. From that time forth her progress toward recovery was satisfactory. Her physical ailments gave way to simple remedies. The power of self-control without any *serious* interruptions returned, and under its carefully persistent cultivation advanced to its natural and sufficient development. On leaving us she said, "I am well, and shall keep so ; thanks, you have taught me the necessity of self-denial. For the first time in my life I have been made to obey."

The same appliances that tend to make life in a well-ordered house beautiful and happy, may be brought to bear upon the disordered mind ; and its wanderings and vagaries be arrested by putting it as nearly as possible in relations like those of private secluded home-life. The great caravansaries we call hotels are not homes, neither do the immense structures we build as hospitals, however well kept, tend to promote the home content, and to awaken those sweet and restorative feelings that belong to home itself.[1]

---

[1] Some years ago, a young lady of about seventeen years of age, of delicate organization, sensitive and nervous, and in frail general health, was brought from a distant city to our care by her only and older sister.

They had been left motherless at an early age. The neglect of rightful early training, the varying impulses of a wayward will, some sad hygienic mistakes—the result of educational neglect,—made her committal to the Retreat a discreet and timely measure.

After the formalities of admission, the sister, after giving the outline of her case, besought me to give some little extra attention to my new patient. " It is a terrible thing, sir," she said, " to leave this poor, sick, dear child, hundreds of miles from home, all alone among strangers." I was ready to promise what she wished, hoping in some way to allay the not unnatural nervous excitement of both.

After expediting their separation, I told the young lady that, being now under my professional care, she could better tell me all about her troubles somewhere out on the lawn. There, surrounded by those soothing influences which I have so often found efficient aids to professional treatment, I at first led her attention to our beautiful surroundings rather than to the details of her ailments, calling her attention to the varied scenery, the pleasant changes of light and shade, and to the flowers of which she seemed so fond ; occasionally alluding to the events of her long journey, and casually asking leading questions about her health,—all this with the pleasant result of gradually substituting self-control and smiles for nervous agitation and tears. The songs of the birds attracted her attention. " I am so fond of music," she said. This I was glad to hear, as we had so many musical parties. Disclaiming the rare musical merit I had heard she possessed, as I called for evidence, it came first in sad, and by and by in merrier, notes. The noon-day bell put an end to

"How clean and nice this room is," said a director to me, one day, in one of the old, rigidly plain halls, long before the reconstruction. "Yes," I answered, "the floor, the bed, the walls are white—if not white as snow —white enough to chill the heart of the delicate, refined young mother who is to occupy it to-day." "Why, what better would you have?" he asked. "All possible home-like ornamentation, neutral tints, pictures, flowers, etc., etc. ; every thing to give the room an inviting aspect, and not painfully to remind her of that refined and *home-felt*

---

our pleasant out-of-door visit. "What is that for?" she asked. "It is noon, and we must go in to get ready for dinner," I told her. "Noon!" she exclaimed, "how strange, how really absurd ! why, a little while ago to be left here all alone, was to break my heart, and now I have been singing and laughing with you." Then came my sought opportunity. "No, this is nothing absurd ; all this is hopeful to me ; you have given me the keynote to the history of your illness ; this gives good promise of your restoration to health and a happy future, if you will only trust yourself to my guidance and accept my teachings." "Why !" she promptly answered, "of course I shall do so." After this I held the case easily in hand. Right heartily did she hold to the faith so cheerfully professed. Her vivacity and kind-heartedness made her a favorite member of our family. In due time, after she had left us as recovered, the evidence of her permanent restoration to health and to new purposes of life coming to us through the echoes of marriage bells, was accepted as the natural sequence of that *morning on the lawn.*

I shall ever remain grateful to those liberal friends on record in the Report of the Retreat for 1861, who gave to my personal solicitation the ample means of developing the natural beauties of the lawn, the erection of the museum, the Ives amusement hall, with many orna-mentations which have since been found to be such efficient thera-peutic remedies.

Medicine is justly divided into "Prophylactic, or the art of pre-serving health," and "Therapeutic, or the art of restoring it." Moral Therapeutics have a wide range ; they are so effectual in al-laying the undue excitability of the disordered mind, and in diverting the current of morbid thought back into the natural channels of health.

After the correction of those physical disorders (so generally ac-

room in which she has left her infant child." Bearing
this in mind, when Mr. Vaux, the architect, came up from
New York to examine the Retreat as to its reconstruction, he asked me, " What would you have me do with
this old building ? " " Reconstruct the ' Lunatic Hospital ' thoroughly, and develop the ' Home ' for the nervous
and insane," was my answer. The result was a success.

In one of my earlier reports of the Retreat I stated
that of one hundred and eighty-seven female patients
last admitted, thirty-four per cent. were wives of farmers

---

cepted as "the punctum saliens," or starting-point, of insanity), I
have often turned successfully to this department of Therapeutics for
that " Medicamentum gratia probatum " of quaint old Paracelsus,
the " remedy approved by grace," for restoring the sufferer from insanity to a right mind.

The whole forenoon I gave this young lady was not wasted. I
know not how it could have been more profitably, humanely, or scientifically employed. I met her afterwards more or less frequently
every day, keeping quiet but careful watch over her progress.

In my regular morning visits to the various departments of the
House, it was my custom to have its immates, who were well enough
to leave their rooms, meet me in the parlor attached to their respec-
tive halls. Of course private and confidential interviews were frequently required. These every-day family-like gatherings proved
far more than formal visits to each room. Such informal meetings,
like many of my old-time visits in general practice, were so homelike
and natural that their influence readily drew out those genial and
social elements of the heart which, in the most extreme developments
of insanity, I have rarely, if ever, found beyond our reach. For here
was possible " that intangible permeation of a household not too
large for the personality at the head of it." The conversations were
easy and natural, often bringing out the mental peculiarities of one
and another ; these in turn leading to timely criticisms, and then to
frank and general discussions.

The best results depended upon the discreet association of varied
elements. Judicious antagonisms worked better than similarities ;
the excited and melancholic amended each other, while the appeal
to the judgment of the majority, confirming the best decision, bene-
fited both. I have again and again traced the final results of successful treatment to such persuasive agencies, by which doubts were

and mechanics—an undue proportion of these classes in
the State. Many were young women, leaving nursing
children at home. In these cases it had naturally followed from the sequences of child-bearing and child-
nursing ; the too frequent entire absence of all or of only
the brief week's service of the "monthly nurse"; the
accumulation of household duties and drudgeries ; narrow and near-sighted economies, and absence of needed
relief of change and recreation ; that the exhausted wife
lost, in due time and course, her appetitite, sleep, and

---

resolved, darkness lightened, excitement subdued, and the better
mind restored. Pleasant as were my visits to such associations, I
often found that I could much more readily decide upon the best
remedy for the excited maniac, than meet to their satisfaction or my
own the sharp questions presented by their ready wit.

The personal influence of the superintendent is naturally trans-
mitted to the assistant physicians, and, by their concurrence, to all
other employés. The best results in such a singularly complicated
household can only be obtained when its various officials act in unity.
Emerson says: "The institution is the shadow of the man." He
must be such a one that the shadow, to be protective, must not be
too thin !

They are greatly in error who think that an attack of insanity
necessarily eclipses all the faculties of the mind. This extremity of
disorder is rarely seen. Frequently the one delusion (sometimes the
many) are kept out of sight so ingeniously as to elude the unskilful
observer. The history of legal investigations, and the experience of
experts in insanity, are full of such curious instances, and not the
least are the illustrations of the ease with which judge and jury are
sometimes misled.

I record one of many cases illustrating that mania in the highest
state of excitement is rarely if ever lost to the power of partial self-
control. A lady once under my care, a person of superior intelligence
and mental ability, belonging to the highest circle of society, became
so furiously insane that she was destructive and dangerous to approach, and obliged to be placed in a room without movable furni-
ture. A member of my family in the habit of making frequent visits
to her, came to the ward to see her. The attendant said : "It is
impossible for you to go to her room this morning ; it is dangerous."
(The lady had a beautiful bright scarlet shawl around her.) "She

general strength, and consequently her self-control, and was compelled to seek the recovery of all in the Retreat. The very large majority of these cases were returned to their families in good physical and mental health ; the avoidable causes of their insanity frankly and plainly explained to them and to their friends, with instructions for their future avoidance. While remedying the marked physical disorders and debilities which, almost without exception, were found to exist, I patiently sought to gain the confidence of each one by the persistent use of individualized moral as well as remedial treatment, gradually winning their sincere trust by sympathy with their condition, seeking to relieve it by pleasant and varied occupation, recreation, and amusement, all evidently for the one result of promoting their restoration to health and home duties.[1]   From this confidence there naturally

---

will tear your shawl to atoms." " Well," she said, " I will risk it." She opened the door and met the patient very cordially, who at once exclaimed, " Oh ! what a lovely shawl ! "   She caught it off, and wrapping it around herself, sat down on the floor with her visitor, and had a long, pleasant, and amusing conversation. The visitor was most reluctantly parted with ; the shawl replaced kindly and carefully by the patient. After her recovery the lady often referred to this visit.   " That shawl was so attractive," she said.

September, 184-.
[1] It is with much pleasure that I take my pen to address you in respect to my dear wife. She stood the journey remarkably well, and glad indeed was she once more to hail her home and friends, especially her little daughter. My wife appears as well as she ever did ; she has a good appetite, sleeps well, and gains flesh every day. Were it not for the deep trouble and anxiety through which we have passed, we should not, from any thing on her part, know any thing of the painful affliction through which she has passed. We are all now gathered in our little circle, and while I write she is singing like a lark. Happy are we, indeed, to have once more so dear a one to make her home happy, and to banish the desolation and loneliness

came out to me the simply told stories of home—cause
and effect—stories that were of lives more patient, unsel-
fish, devoted, and oftentimes tragic than any novelist, but
one,[1] has ever portrayed. I could fill page after page
with these pathetic illustrations of the causes of insanity.
I often found that the instructions given to these patients,
when discharged, and to their friends, prevented the re-
currence of their insanity, The experience of both
parties, together with the earnestness with which they
urged others similarly afflicted to trust themselves to its
curative influences, with the frequent tokens of their
grateful remembrance of benefits conferred, did much to
dispel the too common, yet causeless, dread of the Retreat. See Appendix (2).

If circumstances, in despite of experience and instruction, compelled them to be again exposed to the same
malignant influences, would the consequent relapse and
re-admission to the Retreat invalidate the first record of
" Discharged Recovered " ?[2]

---

that rested like an incubus on all our spirits in her absence. Much
cause have we to be thankful to the God of all our mercies for the
restoration to health of one of the best of wives, the kindliest and
loveliest of mothers, the choicest and most valued of friends. Nor
shall we soon cease to remember you, and the rest of those kind
friends who have been instrumental, through the mercies of Israel's
King, in bringing about so important and desirable a change. May
Heaven shine propitiously on your arduous labors, and crown your
efforts in behalf of suffering humanity with success. We have all
been at my father's to spend the day, and it was a real old-fashioned
gathering I can assure you. There was no sighing, and anxious,
painful longing for the dearest of our flock.

[1] Rose Terry Cooke.

[2] More than all other physical diseases, insanity requires a pro-
tracted period of carefully guarded convalescence. The thorough
restoration of the fractured limb needs the curative help of the splin-
ter or the crutch. The successful physician keeps a sharp watch

A healthy young fisherman, by needless and repeated exposures in the Cove, inducing an attack of acute rheumatism, "hereditary" in his family, is admitted to the Hartford Hospital. He is successfully treated, and in due time discharged recovered. If, in two or three years afterwards, disregarding experience and professional advice, he repeats the exposures, induces a return of rheu-

---

over his convalescing cases of typhus, pneumonia, and diphtheria. The military commander guards sharply against the possible midnight attack of his apparently defeated enemy. The prolonged residence at the Retreat has often proved its practical wisdom by the results of confirmed convalescence. I recall one of many illustrations.

A farmer, fairly well to do, in middle life, of a nervous temperament, hardly up to the standard of general health, was brought to the Retreat in a state of high nervous excitement. The death of his father had left a considerable property to be divided between himself and other heirs. An acceptable division had been agreed upon, excepting the right to an outlying piece of land, on which was a valuable barn. The quarrel about this upset the reason of this poor man, and brought him to the Retreat. In reasonable time his debilitated system yielded to treatment, the effect of unwise discussions and needless excitement giving way to quiet, genial influences and pleasant surroundings ; all which, with the aid of good digestion and sufficient sleep, led him to accept the evidences of his mistake regarding the disputed property. This had been fully explained to me, for my possibly timely application. In this state of early convalescence, his brother came to see him, finding him out-of-doors, quite at liberty, cheerful and reasonable, and apparently "as well as ever." The brother insisted upon taking him home, where he said he was greatly needed. My advice was overlooked, as my apprehension was deemed unfounded. The result confirmed my judgment. The man was brought back the next day in a highly excited condition. The brother reported that he seemed as well as ever while they were returning home, till, unluckily, to shorten distance, they took a cross-road that brought them suddenly up to the old barn, "when he jumped right up in the wagon, as crazy as ever." He seemed glad to be back at the Retreat, remained willingly, and finally recovered. On his next departure, bidding him a hopeful good-by, I jocosely told him not to take that shorter road home. "Never you fear, doctor," he replied, "it will be a long time before I go near that confounded old barn again." As he was never re-admitted I am sure he found that the longer way round proved the safer way home !

matism, is again admitted, and again discharged recovered, does the last record annul the primary one of recovery ? and if not, why should not the same law of reported results, valid at the Hospital, obtain across the street, at the Retreat for the insane ?

An eminent writer (Huxley) points out that—

In the present rapid growth of the minutest branches of most of the sciences, and the consequent tendency to narrowness which this diminishing scale of research seems likely to invoke, all men of science should be primarily so educated as to secure breadth of scientific education without superficiality of knowledge, as the best security against the natural danger of drifting into narrow specialties.

In view of the largely increasing varieties of disorders from which insanity may originate, and of the many new remedies, in addition to the better knowledge of the old, which come to the aid of medicine in the progressing art of preserving and restoring health—the broader reachings of prophylactic and therapeutic agencies—the practical alienist should possess not only a readily available knowledge of all this, but keep a careful watch over the possibilities of the future. The comparison of the Dispensary (or " Bigelow's Sequel ") at the date of my graduation, in 1828, with that of to-day, naturally suggests the possible future of some of the one hundred thousand weeds " whose virtues," Emerson says, " are yet to be discovered." Certainly most, if not all, the original thirteen members of the Association had a large experience in general practice before assuming special charge of the insane.

Dr. Parkes says : " Hygiene aims at rendering growth more perfect, decay less rapid, life more vigorous, death

more remote." The acceptance of this art or science of
preserving health has within the past thirty years added
nearly four years to the average life of the men and
women in England. The humane and scientific re-
searches of Dr. Bowditch, of Boston, have largely limited
the ravages of consumption in New England. The establishment of thirty-one State Boards of health, in the
United States, since that of Masschusetts, in 1869, and
the increasing acceptance of the vital necessities of sanitary reform, all lead us to hope that erelong consumption, malaria, and insanity, with many other formidable
enemies of health and life, may be met by some new an-
tagonistic power.

In measuring our means of arresting insanity we must
accept the science of prevention as a higher power than
the science of remedy, a power to be looked for outside
the wards of a hospital. "True medicine," says Dr.
Richardson, "now stands boldly forth to declare the
higher philosophy—the *prevention* of disease." "Our
art," says Dr. Bowditch, "looks still higher, to the prevention as well as the cure of disease." Prevention
justly takes precedence. Very many of the ordinary
causes of insanity may be easily avoided, and, if needlessly induced, may be readily overcome. They are the
natural outgrowth of heedless or ignorant violation of
well-established laws of hygiene—laws that should be intelligently taught in every common school in the land.
There are other causes of far graver import than they at
first suggest, where prevention demands that the earliest
symptoms should be promptly recognized and efficiently
treated. Dr. Tuke says :

The prevention of disease is the first and most earnest intention of medical science in all its departments. The prevention of mental disease is clearly within the scope of the physician's highest aim.

He further says :

No medical forethought can prevent the occurrence of insanity from accidental causes, but a vast proportion of the insane become so in consequence of physical conditions of life and modes of living, which lead to the result as certainly as unsanitary conditions of physical life lead to typhoid fever or tuberculosis. It is in such cases that a prophylaxis can sometimes be established. Moral treatment is the true prophylaxis. If the most favorable instances of these ailing minds are brought under the influence of strong and healthy minds, the fearful heritage may oftentimes be avoided.

Dr. Conolly dwells at length upon the effects he has witnessed from the " individualized treatment "—the influence of a sane, addressed to an insane mind.[1]

---

[1] Universally in zymotic disease, and generally in all others, the practitioner finds a law of rise, progress, and limitation. This accepted, it remains to watch over the course of the disease, antagonizing the deviations and the often perilous complications. No such law is found in insanity ; its origin is more obscure, its range is wider, its development is more varied, unexpected, and seemingly unreasonable. Sometimes in its formative stage, a specific delusion originating in a doubt or question develops gradually, growing with the growth, and strengthening with the strength of the disorder, until it becomes a fixed habit ; the mind running in a rut influences every motive and thought of the daily life. This will sometimes remain painfully dominant after every other symptom has yielded. Few conditions of a patient under treatment are more painfully embarrassing ; the apprehension naturally arises that some one portion of the brain, and a very limited one it may be, is seriously, perhaps hopelessly disordered. Here the natural hopefulness of trained individuality leads to persevering effort, in the hope that some sharp, unexpected impulse may shake off this incubus, which like Sindbad's " Old Man of the Sea " threatens to hold on to the end.

Some years ago a gentleman about thirty years of age was brought to the Retreat in a truly deplorable condition. He was an officer in

In the application of moral treatment it is of vital importance so carefully to scrutinize the environment of each patient as to avoid as far as possible all depressing or exciting influences. All accept the axiom that cheerfulness and sympathy in any sick-room promote the best working of remedies ; in the wards of a lunatic hospital, remedies avail little without their coöperation. I have found few things more depressing and harmful to the recent and hopefully curable cases of insanity, than even the sight, and, much more, the association with the demented and hopeless. To such cases (excluding the few the severity of whose disorder prevents any realization of their condition), finding themselves in the bewilderment of such surroundings, their natural conclusion is : "I am one of a hopeless crowd ; what better chance can

---

active service under the national government. His duties, involving large responsibilities, under harassing circumstances, in a tropical climate, were prolonged without change until he broke down both in body and mind. A rigid sense of duty had led him without complaint or request for relief to this possibly fatal overwork. His case presented on admission an unusual combination of discouraging symptoms. Happily his one overruling delusion disposed him cheerfully to accept his position with us. After some months of careful watching, patient care, and that due medication which the condition of nearly every physical organ demanded, his general health seemed fairly restored. He had a good digestion, and plenty of that "best food of the brain," sleep. Notwithstanding all this one great anxiety remained to me in his delusion, the vividness and tenacity of which I have rarely, if ever, seen equalled under similar conditions of general health. No belief whatever was more firmly held by him than that for a neglect of official duty his life was forfeited ; the Department having resolved to order his execution in a signal and impressive manner. He confessed the offence was a trifling matter,— in truth the oversight of a subordinate. But its frequent repetition, in contempt of repeated orders, had compelled the Secretary to make an example. All reasoning with my patient seemed worse than useless ; the only comfort I gave him was the assertion that no government, State or General, could take him from the Retreat against my

I have of recovery?" The number of patients in some
of our State hospitals exceeds the population of more
than each one of forty of the towns of the State of Connecticut. In several the recent case on admission becomes the legally committed citizen of a community
whose annual official report records more discharges by
death than by recovery from insanity. Classified, however carefully, as the multitude may be, the different individuals must come frequently in contact in the chapel,
and in the means of their recreation and amusement.
With such immediate surroundings the recent case can
hardly look from his window or step out of his door without seeing or hearing some hopeless victim of a disease
from which he has fainting hopes of his own recovery.
Reason as you may with him, for the present time, at

---

record of " not recovered." Afterwards. avoiding the discussion,
I continued every effort to strengthen his general health, and to
amuse and occupy his mind, awaiting my opportunity. and this in
time came. One morning, when he seemed more bright and genial
than usual, I referred jocosely to our old difference of opinion, and
confessing that discussions were useless with one who clung to a delusion with the obstinacy of Tam O'Shanter's wife, who " hugged
her wrath to keep it warm," I told him that I was sure that in good
time his large stock of common-sense would come to the front, and
added most seriously and earnestly : " If I were now to give you the
endorsed testimony of your Department of your deserved high standing as an officer, and that your official record was without fault or re-
proach, I really fear you would reject it." " Oh ! no indeed, I
would not ! if you would only show me that," he replied. Drawing
from my pocket the formal official document, I answered : " I re-
joice that I can do so. Here, in reply to my letter of inquiry, is the
prompt and cordial answer of the Department, a flattering certifica-
tion of your high position in the service." He received it with
much emotion. The effect was electrical. He seemed to awake, as
it were, from an oppressive dream. When he came down next
morning to make arrangements for his return home, he was a very
happy and sane man.

least, the "twirling thumbs" will beat down your sanitary arguments.

We believe that absolute segregation is possible, and is consistent with a large economy in construction and in current expenses. The telegraph, the telephone, and the tramway may bring the annex of the main hospital sufficiently near, so that out of sight and out of hearing, at the distance, more or less, of a mile, these apparently separate institutions are within easy reach of the sharp supervision of the chief superintendent. Again, the query may be raised whether, in the continued growth of very large institutions, there may not be developed in the future a school of hospital specialists, simply executive officers, skilled in economic management and training, instead of broad, earnestly sympathetic, and versatile physicians of large experience, through study of individual cases.

Our position is amply confirmed by authorities who have the right to speak and to be heard. At the meeting of the International Medical Congress in Philadelphia, in 1876, Dr. Ray said :

As the result of my own observation and experience I am convinced that four hospitals of three hundred patients each can be both built and maintained at a less cost than one of twelve hundred patients, equal provision being made in both cases for the kind of care to which the insane, even in the lowest grades of the disease, are entitled.

Again he says :

I doubt whether it is possible to have in these mammoth establishments certain qualities of administration indispensable to their highest purposes. The animating spirit, the close, thorough supervision, inspiring, guiding, correcting every movement, and essential to our highest ideas of hospital management, will be but feebly maintained under such conditions. The patient is but an atom in the great mass around him, losing the attributes of humanity, sane and insane, in the technical character of patients.

At the same meeting Dr. Kirkbride said :

It is fully shown by reliable statistics, as I believe, that the people of the State will derive more benefit from several small hospitals in different parts of the State than from one large one at a central point. And I think it will also be found that the former can be provided with quite as small an expenditure of money, and could be carried on at no greater cost per patient. . . . There is one advantage in these smaller hospitals I cannot avoid referring to, and that is the personal intercourse which a superintendent is able to give to his patients when their number is not so great as to prevent his paying daily, or very nearly daily, visits to each. I believe this to be one of the most important of all his duties, and one which, certainly, if he is rightly constituted for his position, *no one can do for him.*

Dr. E. C. Seguin, of New York, in a letter to a member of our Legislature, 1880, says :

The vast majority of the insane are afflicted with chronic and incurable disease. They need only humane treatment, the largest possible amount of personal liberty, plenty of occupation, some amusement, the plainest quarters, and the simplest diet consistent with the demands of modern hygiene. They do not require the attention of as high a grade of medical talent or as numerous and skilled attendants as do acute cases. They can, I believe, be well taken care of at a comparatively small cost. It seems a reckless waste of money to build palatial hospitals to be filled with incurables. . . . The curable insane need the highest medical skill which a large salary can attract ; a much larger number proportionately of assistant physicians selected by severe examination ; many real nurses, not mere attendants or guardians. They require the best food, with the liberal use of costly medicine, etc. It is economical and humane to spend money freely in order to facilitate recovery.

An intimate knowledge of the condition of the insane for more than half a century has given me an increasing sympathy for them as the most grievously afflicted of the human family. Both my experience and observation since the meeting of the Association in 1844 have convinced me that the great and unexpected changes in their numbers and relative condition demand some modification of the original propositions. That which gave two hundred as the preferable maximum of patients to be treated in one building, may again be as wisely accepted in connection with a new classification. See Appendix (3).

In conclusion, let me quote from Dr. Ray, one known to us all, and best esteemed and honored by those who knew him best, and who, in his description of the " Good Superintendent," has given us a most happy photograph of his own hospital life :

The " Good Superintendent " constantly striveth to learn what is passing in the mind of his patient, by conversation and inquiry of those who see him in his unguarded moments. He also maketh diligent inquiry respecting the bodily and mental traits of his kindred, knowing full well that the sufferer is generally more beholden to them than to himself, for the evil that has fallen upon him. He endeavoreth so to limit the number committed to his care as to obtain a personal knowledge of every wandering spirit in his keeping. He boasteth not of the multitudes borne on his register, but rather, if he boasteth at all, of the many whose experience he has discovered, whose needs he has striven to supply, whose moods, fancies, and impulses he has steadily watched. To fix his hold on the confidence and good-will of his patients he spareth no effort, though it may consume his time and tax his patience or encroach seemingly on the dignity of his office. A formal walk through the wards and the ordering of a few drugs compriseth but a small part of his means of restoring the troubled mind. To prepare for this work and to make other work effectual, he carefully studieth the mental movements of his patients. He never grudgeth the moments spent in quiet, familiar intercourse with them, for thereby he gaineth many glimpses of their inner life that may help him in their treatment. Among them are many sensible to manifestations of interest and good-will, and the good physician esteemeth it one of the felicities of his lot that he is able to witness their healing influence. He maketh himself the centre of their system, around which they all revolve, being held in their places by the attraction of respect and confidence. To promote the great purpose of his calling he availeth himself of all his stores of knowledge, that he may converse with his patients on matters most interesting to them, and thereby establish with them a friendly relation. The unwelcome communication he ever tempereth with soft and pleasant words, thereby verifying in himself that saying respecting a worthy of old, that he made a flat refusal more agreeable than others did the most thorough compliance.

In my report of the Retreat for 1860, I remarked that over three thousand cases of insanity have now come under my direct care and observation. In a large proportion of those cases whose history I could obtain, I have found that the remote and predisposing causes of insanity could be plainly traced to the malign influences of childhood.

In this connection I quoted the following from General Oliver's report to the Massachusetts Board of Education :

While we abundantly provide for the thorough training of the mind, we almost wholly neglect the training of the body, and the effect of this pressure upon the intellect without corresponding pressure of the body is that the latter suffers, and by degrees the feebleness which is generated by this want of proper physical exercise of the body extends to the mind ; for the twain are in incomprehensible mystery of connection, and each is participant of the other's strength or weakness. So then the mind becomes less vigorous by reason of the fading vigor of the body, as the body is always weakened by the fading powers of the mind, and each gradually participating in a gradual antagonism to the efforts of educators and the efforts of self-education. This is especially true of our girls. Our boys indulge more in vigorous and active exercises. Athletic sports are full of interest to them, and into them they go with a rush and a relish and a heartiness of fun most cheering to behold, and most excellent in its influence upon their bodily health. But of how little physical exercise do our girls partake, and how quick are we to check any propensity to activity in play and to any romping gambols or vigorous recreation on their part. . . . I venture to say that not more than one girl in ten nowadays enjoys real sound, rugged health, and surely that is a very unwelcome statement about those who are expected hereafter to be helpmates to husbands and mothers of children. . . .

In an admirable article upon Insanity and Hospitals for the Insane, prepared for the National Almanac, Dr. Earle remarks :

That it is not the regular employments of mankind which are the most prolific causes of insanity. It is rather those habits, customs, and other influences which minister to his appetites, stimulate his passions, and most powerfully operate upon his sentiments.

Intemperance of all kinds, debauchery, self-abuse, all high popular excitements whatever may be the subject, these excite and exhaust the nervous energy ; and grief, anxiety, troubles, difficulties, and disappointments greatly depress it. To these influences then we may rightfully look as among the most powerfully exciting causes of the disorder in question.

In the thirty-ninth of the Retreat's reports it will be seen that of nine thousand four hundred and seventy-three cases, as given by Dr. Earle, being the total of all cases admitted in four prominent hospitals wherein the causes of insanity were given, seven thousand five hundred and ninety-one, or four fifths of the whole, were the results of some one of ten causes, all of which were such as exhaust, debilitate, or depress the vital or nervous energies. A sensual and selfish, or idle and aimless life, must inevitably act as a predisposing cause to the development of one or more of these causes. In a large proportion of the cases which have come into my care insanity might have been prevented by the use of well-known measures, or natural and right development of body and mind, wise aims in life, and a reasonable exercise of self-control. The power of the will to control the insane impulse is great, but its power to effect this result must be trained and be made conscious of its supremacy. The question, therefore, how shall I escape insanity? is one capable of a more direct and explicit answer than many parents and educators of youth seem to imagine. See Appendix (4).

It follows, as a necessary consequence, that whatever makes us better or wiser, gives us more correct views of our duties to God and our neighbor, and at the same time gives us more courage, strength, and willingness to do that duty, places us so much more beyond the reach of these causes of insanity, and gives us also the greater ability to resist successfully the attacks of this disease when induced by causes beyond our control.

Insanity, as a strictly physical disease, comes eminently within the range of preventive medicine. When our proposed and thorough system of State sanitary registration in Connecticut is carried out (if ever), and each

case is reported in its earlier stages, we may hope to attain a more accurate knowledge of the predisposing and
exciting causes of this malady, which is filling our lunatic hospitals faster than we do or can build them. We
can also more efficiently apply the means of prevention
and remedy, when we can better measure its various
causes : erroneous educational and social influences, neglect of family training to reverence and obedience, sen-
sational reading, evil habits of body and mind, idle, aimless, or sensual life, and learn more exactly, as we shall
learn, how very early in life the predisposing causes of
insanity are implanted in the child.

During the present century, no greater progress has
been made in any department of philanthropy and science than in the direction of the better care and treatment of the insane. A greater work remains to be done,
a work greater than cure or kindly care—that of prevention ; a work which, in order to be of the highest success, must reach back often to the early life, the family,
the nursery, and the school.

The question before us to-day is not only, what can
the State do for the chronic insane ? but the wiser and
more timely question, how can we prevent insanity ?

The neglect of physical training, and the imperfect
physical development which follows from this neglect,
were strikingly evident in many of my female patients.
The various causes which were reported to me as the
sources of disease, and which are classified in the tables
under the head of " ill-health," " undue mental effort,"
" grief," " domestic unhappiness," etc., could very frequently be traced, in their primary influences, to the one

cause of a want of physical stamina. We press the training of the mind, by all possible hours of study in and out of school, and by the added stimulus of emulation, while we neglect the training of the body, in disregard of that mysterious but absolute law of sympathy, which compels the debility of the latter to cripple the action of the former. In the same line, is the prevention of excitement so happily illustrated in the Northampton Hospital by Dr. Earle, where the busy day on his thoroughly cultivated farm, followed by a quiet night, illustrates, happily, the sanitary results of wisely directed occupation.

Life has been compared to a line—that of birth, the point of origin, that of death, the point of termination, the length of the line between being an uncertain quantity under a supposed secret and inexorable law, over which we were ignorantly believed to have no control. The history of the human race has ever testified to the incessant craving of the heart that "our days may be prolonged in the land." The Science of Preventive Medicine justifies this innate desire by demonstrating that it possesses the power to give a longer extension and a more definite and certain quantity to this "line of life." We are told that the days of our years are threescore years and ten, and that if we are deprived of the "residue of our years," and do so generally fall far short of that attainment, it will be well for us more carefully to regard that wonderfully true and perfect sanitary code given to the Jewish nation, and recorded for our instruction and guidance in the Holy Scriptures, and remember through their obedience to those hygienic laws " He in-

creased the people greatly, and made them stronger than
their enemies," and when he brought them forth out of
the land of Egypt, "there was not one feeble person
among their tribes." Nor one insane !

In many of the insane the power of observation is active and the understanding has a considerable range of
exercise, while the affections exist as warmly and the
sensibility is as acute as in a state of perfect mental
health. The utmost care therefore, should be taken to
act on what remains of intellect, wisely to direct the impaired faculties of the understanding, and at the same
time to cherish and govern the affections by all the re-
sources of compassionate protection.

To suppose that the inmates of a lunatic asylum must
necessarily be in a state of continual unhappiness, is as
erroneous as to suppose the asylum itself a place only
for confinement and suffering. There is a wonderful
diversity in the manifestations of this disease, each case
having its peculiar character : the melancholic, who sup-
poses God has forsaken him for time and eternity ;
the excitable, defying all law but his own sovereign will ;
and those who, less gravely affected, are yet unfitted for
the duties of life, and are waiting, with more or less of
patience and resignation, the time of their recovery.[1]

---

[1] The history of one of my patients in the Retreat made such an impression that I cannot refrain from describing it in some detail. She
was a married lady, under middle age, well developed physically,
with a genial and easily impressible temperament, which, with her
happy manner, made her a universal favorite. On admission to the
Retreat she left an intelligent and devoted husband, several charming children, and ample means in a pleasant home. The history of
the case, as given to me, showed that the gradually widening scope of
maternal duties and household cares had unconsciously led her to a
devoted but really reckless draft upon her physical energies, which

Now, as amusement and recreation are essential to the preservation of the health of body and mind, and as their genial influence is fully appreciated by us during the convalescence from an ordinary illness, how much more sensitive to their effect must be those who are suffering under this graver disorder.

---

in due course brought on a disorder of the widest range of the digestive apparatus, and consequently and most naturally of other equally vital organs sympathizing with it. The derangement of the nervous system was the common and natural sequence of loss of appetite, loss of sleep, and loss of both physical and nervous power.

I found her emaciated, enfeebled, and dejected ; the whole system "fagged" (I can find no better word for its best description). The case was one of functional disorder, primarily induced by readily preventible causes ; easily remedied in its early stages by simple means. But the system had literally drifted into a condition where there was, at first sight, some apparent reason to fear *disease* of the brain, or change of cerebral structure.

A close scrutiny into the case revealed its simple character. The suicidal tendency which at first I had reason to fear (but which was never discussed between us) soon ceased to worry me. The general treatment required for its ultimate and entire succcess some months of uninterrupted residence at the Retreat ; in the earlier stages the use of ordinary alteratives, sub-tonics, and milder sedatives, etc. ; throughout all, persistent rest both of mind and body ; relief from all duties and worries, with all possible cheerful surroundings and heart-cheering influences ; diversions, in-doors and out-of-doors ; in brief, those social and genial remedies which I have elsewhere more minutely described, and in which my prolonged experience has increased my confidence. My acquaintance with this interesting patient ceased with my record of "Discharged recovered." Of this I am confident : with a reasonable regard to prophylactics in her future life, and fair play given to her naturally large power of self-control, and in all a common-sense profiting from the sad lesson she had learned by heart, there would not be a relapse in the record of her case.

If the love of God, faith in Christ, and the hope of heaven give way under the delusions of insanity to hopeless despair and the certainty of endless doom, suicide seems a natural and logical sequence.

The peril which lurks beneath such cases demands unwearied vigilance. This was fairly illustrated by another patient who was one of a class before described. She was a woman of more than ordinary intelligence and self-control ; she was broken down by the

While, to the insane, all their delusions are as real as
they are, in truth, imaginary, none but those who are in
constant converse with them can realize how material is
the result of an intimate personal intercourse. I daily
saw some cloud brightened, some terror banished, some
wearisome burden lightened, by a few words of advice,

"accumulations of household duties and drudgeries, of narrow and
near-sighted economies," and was brought to the Retreat by friends
in the faint hope of her recovery.   It was an extreme case of melancholy, demanding constant watchfulness.   In due time the apparent
improvement in all her symptoms gave me comforting promise of
her speedy recovery and return to her home, where she was greatly
needed.   My family residence was on the lawn at a little distance
from the Institution.   Parties of our convalescent patients were our
frequent social visitors.   She joined the party one evening, and this
her first visit appeared to be greatly enjoyed ; no one of all seemed
more truly convalescent.   As the party was about returning to the
Retreat, her hostess, learning that a storm had suddenly come up,
and seeing her need of warmer wraps, sent up for a shawl, and
wrapped it carefully round her, adding some brief words.   Her
health was soon confirmed, and she returned home.   A year or two
afterwards, meeting her hostess in the city, she ran up to her with
many expressions of grateful remembrance.   " Excuse me, I do not
remember you," was the answer.   " Oh ! I don't wonder at that ; I
am so changed since I saw you.   Don't you remember inviting me
to your house one evening, and, as we were about to return to the
Retreat, being told that a heavy storm had come up, you brought
down a beautiful shawl and wrapped it carefully round me.   I can
never forget it ; let me tell you—before coming over to your house I
had made up my mind to commit suicide that very night.   I had for
a long time believed that God had blotted me out of the book of his
remembrance, and the quicker I met my fate the better ; that my
husband would be happier with another wife ; my children with another mother.   I knew I was watched pretty closely all the time, but
I had fixed it so that I should certainly have succeeded.   After I had
gone back to the Retreat, thinking over what a nice time I had had
at your house, all at once it came over me.   'What *can* all this
mean ? this lady to invite me, treat me so kindly, and then wrap her
beautiful shawl so carefully about me, lest I should catch cold and be
sick, why God cannot have forsaken me ; I can't be such a sinner ;
how foolish, how really crazy I must have been to think so !' From
that moment I was right.   And now ! I am so happy ! "

of cheer, of consolation, or of sympathy. That asylum for the insane is poorly cared for where the wants of the body are alone abundantly supplied, while the cravings of the heart are left unappeased. Far better, in my view, to banish all other remedies from the wards of such an asylum, than to leave them destitute of that practical,

In connection with the two preceding cases, I here may well introduce the partial history of one more—another vivid illustration of this, the saddest of all forms of mental disease—melancholia. Unfortunately, the writer of the following letter, immediately after its reception, was placed outside of my aid, but never outside of my deepest sympathy :

" DR. BUTLER :—Will you allow yourself to be interested in a stranger who has been attracted towards you by some remarks in an article on the treatment of insanity ? Long years ago (nearly twenty) a deep dark cloud settled on my spirit. I cannot find the daylight ; midnight envelopes me ; a weight of suffering oppresses me. The sense of being is pain. I can conceive of no circumstance that could make existence precious, an habitual blessing, I seem to be in a dreary dream, from which I cannot awake. I move about among men, but not of them ; nothing makes a deep, abiding impression, becoming part of my nature, and arousing all my soul. Things come and go, and almost every thing is intangible. I am not a young romantic girl ; I am over thirty years of age, but alas ! I know not *how to live ;* I die a living death. What can furnish interest, motion, object, to fill the abyss of the human soul ?

" What am I ?
An infant crying in the night,
An infant crying for the light,
And with no language but a cry.'

" It seems sometimes as if I was almost beside myself. I have such a profound sense of want of adaptation to any thing. I am a stranger in a strange land ; I cannot do life's work ; my sinews are cut ; my hands drop at my side. I am a poor crushed spirit, staggering sometimes under the load of life. Of course I have not suffered this without looking upward ; but all in that direction is total darkness. The whole realm of Christianity is a terra incognita to me ; my knowledge of the system is simply historical. Sermons, reading, the instructions of Christians, are all unavailing. My deepest prayer is an agonized *cry for help* to the ' Unknown God.' My religious consciousness has never in my

personal sympathy, whose hearty sincerity so directly tends to the larger development of hopefulness and self-control.

In some varieties of this physical disease (insanity) some articles of the *materia medica* are, in my opinion, essential to speedy or to permanent cure ; in many more,

---

life been awakened.   I have not the slightest realization of the relation between the finite spirit and the Infinite,—my soul and its Creator. When I say : ' Is there a great Being who knows all I suffer, who cares for me, and who is deterred by adequate reasons from coming to my aid?' and receive this reply : ' There is a great Being who is deterred from helping you, till you come to *Christ*,' I feel utterly powerless.   I can utter a cry of vague, undefined want ; but when I am bidden to ' *believe*,' which means to exercise towards Christ adoring, trusting affection, I know nothing about the matter ; I can see nothing, do nothing.   If you put a stone on the head of a plant looking up for the free air and sunlight, if you do not crush it out of existence do you not make it a living tomb?   Is there vitality for the human soul?   Is there a world of life and activity and pleasurable existence?   Shall I ever breathe its atmosphere?   Must to ' suffer and be still' be my life's work?   Must I go on, crushed, paralyzed, benumbed, till body and spirit separate?   If God ' is not willing that any should perish,' must I die to gain my first conception of happiness?   Professionally, is not this an abnormal condition? What is the matter with me? Is there any hope for me? Am I not the victim of mental disease? Do I not speak simple truth when I say I have no power to rise, throw off this nightmare, enter into life's labor, respond to its relations, meet its obligations, and enjoy existence ?

" I am not well ; my back is not strong ; my system wants healthful tone and vigor ; I have habitual dyspepsia ; it is an uneasiness and lack of vitality of the stomach and want of digestive power, but not habitually attended with constipation.   I remember that some such derangements of health were felt soon after my soul sunk down oppressed by this inner suffering.

" And now, have I touched your sympathy for a deeply tried fellow-being ? Can you do any thing for me in the way of suggestion ? Is it at all probable that I shall ever be, do, enjoy, any thing as others may ? Will I live on and on in the darkness and pain, do you suppose, and when from old age or disease ' the weary wheels of life stand still,' must I fear that I will close my eyes on this dreary ante-chamber of being, and ' take a leap in the dark ' ? "

they are useful, soothing, pleasant adjuvants ; but these moral means are so pleasant in the using, they so soothe the heart weary with long waiting for health and home, banishing, for a time at least, those delusions which make the worse appear the better reason.   I claim that

---

A lady, the daughter of a merchant, married, and a connection in business was formed between the father and the husband.

In a short time the embarrassment of the former involved the whole fortune of the latter ; and in about a year the young couple were left without any provision, with one child and the expectation of another.   What added to her affliction was, the trouble of her parents and the other children, for all of whom she had the tenderest affection.   I knew this lady from her childhood.   She never had a good constitution, but had always been subject to severe headaches and other corporeal ailments.   A near and dear relative with whom she corresponded, in the attempt to console, very vehemently exhorted her to seek consolation in religion, which advice she enforced by such spiritual arguments as she thought necessary.   Unfortunately, these arguments were intermixed with many abstract doctrinal points which were new to the sufferer.   In the adaptation of them to her own case she felt great perplexity.   Instead, therefore, of deriving consolation, she at last adopted, without due examination, the most dangerous sophisms for truths.   It was soon perceived that her reason was wavering ; shortly complete insanity was developed.   In this state she was brought to London, and consigned to my direction. She was then only twenty-four years of age.   There was evidently great constitutional as well as mental disorder.   In a few months I had the satisfaction to see her health much improved, and every illusion by degrees vanish.   In a few weeks she returned to the bosom of her family.

Never, probably, has any one who had been insane been exposed to greater risk of relapse ; yet after the first struggle, and experiencing some threatening symptoms, she rallied.

Then it was she experienced real consolation from religion.   Her recent spiritual delusions had passed away.   If she remembered the new lights which had so fatally misled her, and finally absorbed her reasoning faculties, she was aware of their dangerous effect ; and relying solely on those principles from which she had formerly always derived satisfaction and support, she was enabled to preserve her reason and attain a state of comparative happiness.

The above is given by Dr. Burrows of London, in Barlow's "On Man's Power over Himself to Prevent or Control Insanity," page 92.

both of these remedies are essential to the best curative treatment.

Amid the weary hours of sad or fearful imagining, music, games, all social or intellectural gatherings and recreations, excursions, changes of scene and localities, art, in its various forms of beauty, pictures, engravings, statuary, and, above all other things, flowers—they are ever most welcome.

Dr. Poole, of the Montrose Asylum, says :

> After the obliteration of reason, many of the highest feelings of our nature remain, to which a successful appeal may be made, and those by which we are connected with a higher sphere of existence, admit as readily of being awakened, on the proper object being presented to them, as the ordinary passions under which the lunatic acts. Their influence is, in the highest degree, consoling, and congenial to the return to mental strength and serenity ; the effects in each individual are probably as different as in the members of an ordinary congregation.

I cannot forbear quoting the testimony of the Rev. Dr. Gallaudet, for many years chaplain of the Retreat :

> How many torpid sensibilities have I seen awakened to respond to the impressions of the fair, the beautiful, and the good ; how many consciences aroused to a sense of the right and the wrong, so as to produce the power of self-control and of proper conduct ; how many slumbering domestic and social affections kindled up into their former activity ; how many religious despondencies, sometimes deepening into despair, changed into the serenity of Christian hope ; how many suicidal designs forever abandoned, because life had become a pleasure, instead of a burden too heavy to be borne ; how many prayers revived at the altars of private and public devotion ; how many kindly charities of the soul breathing forth, once more, in deeds of self-denying benevolence !

Amid the vestiges of reason, the affections and sensibilities sometimes exist as warmly and as acute as ever, and, in many cases, the same high and ennobling results may be attained as from the operation of similar causes upon individuals under ordinary circumstances. Leaving out of the estimate all other results, my fifty years' experience, thirty-three as superintendent, have confirmed the opinion, early expressed, of the benefits of these influences as remedial agents. Any deviation from good order and propriety, during chapel service, has been no more frequent than interruptions from impatient and undisciplined children during " meeting " in the country.[1]

---

[1] During the Rev. Mr. Gallaudet's services as our chaplain, a very insane woman was admitted. She was the wife of a well-to-do farmer, and was of more than usual intelligence, of kindly, cheerful, temperament, and naturally of large self-control. She had broken down under the too common influences of monotonous overwork and worry. I found her general health seriously impaired. She was excitable, impulsive, indeed utterly out of reach of control. Her language was profane and obscene beyond all precedent in the somewhat large observation which my varied fields of practice had given me. I placed her in rigid seclusion, not allowing the matron or the nurses in my visit to be exposed to such foulness of words. In due time her general physical health decidedly improved. She had a good appetite, and slept well ; but neither persuasion nor reproof could tame that unruly member, the tongue. My hopes of speedy recovery were fading, when one afternoon, just before the chapel service, her nurse, passing through my office, remarked—not as a request to be heeded, but as something strange : " Mrs. B. wants to go to prayers." Recalling her as she passed out, I asked : " Was the request made without any suggestion ? Does she really want to go ? If so, take her up with you. Sit near the door, and watch her sharply. If she says a word, hurry her out." A few minutes' reflection convinced me that this time at least the first conclusion was not a wise one. A few evil words from the poor insane woman would disgust many with the chapel services ever after. But as the bell was ringing it was too late to revoke the order. On going to the chapel I found her seated near the door. She gave quiet attention to the whole service, and peaceably retired. Telling her the next morning that I was pleased to see how quiet she was in the chapel, she answered : " The nurse was afraid to have me go." " Do you wonder," I said, " talking so

(I must remark that it is self-evident that the chaplain should be like the assistant physician, the direct appointee of the superintendent).

We are largely indebted to Dr. Dewey, of Kankakee, Ill., for a most timely article in the *Alienist and Neurologist* of January, 1884, giving an exposition of the rise and progress of the separate systems of " Congregate " and " Segregate " buildings for the insane, *i. e.*, the connection or separation of the different classes. In a brief recapitulation of the lines of thought in general, he says :

1.   Institutions for the insane were at first only founded for public relief, and without the idea of benefit to the insane.

2.   It has always been a too general impression that the insane are essentially different from the sane in every thing, instead of the fact being recognized that they possess natural traits and activities, which are, however, modified through the agency of disease, wrongly directed or held in abeyance ; and this mistake has been very mischievous in its effects upon the provision for them, preventing a supply of a natural and domestic abode, adapted to the varying severity of different degrees and kinds of insanity.

---

badly as you have done ?" "Oh ! doctor," she said, "you never need fear that. I know better than to use unholy words in God's holy house." I "improved the occasion " by leading her thoughts back : " What would your good friends at home think of you," I said, " to hear you talk as you do—your good minister's wife, and the deacons, your husband, your own children, and all the good folks who, you say, think so much of you ?" This appeal subdued her, hopefully ! " Now," I added, "if you behave like a lady in God's house, you can do so everywhere else. You can do so with those good folks upstairs who long to have you with them, and in a nicer room than this, and away from these poor unhappy people whom you can't like !" Asking it also as a great favor to myself, I won the promise of better behavior, and only good words. Bating some few incidental and brief relapses, her promises were faithfully kept. After a few weeks of pleasant and uninterrupted association with quiet and convalescing patients, her restoration to health, and her return home was a cheering illustration of the principle we are advocating.

3. The essential difference between an institution for the insane and all other institutions, in confining and controlling those who are held as prisoners without being guilty of any offence, and who are entitled to the utmost privileges and consideration of their wants, without possessing in the eye of the law or in the exercise of reason the ability to enforce their claims, was long overlooked, but has come to be more fully appreciated.

4. Gradually insanity has come to be recognized as a disease, hospitals have been founded, mainly for curative treatment, and the congregate asylum has been developed, admirable for its purpose, but not adapted for universal application to the entire body of the insane.

5. Finally, the infinite variations among the insane, in the manifold forms of the disease ; in the degree of reason and self-control possessed by different individuals or characterizing different groups of the insane as a whole ; in the various classes of private and pauper, criminal and innocent, epileptic, inebriate, etc., are beginning to be more fully understood by the public and the medical profession, and a variety is being introduced in the erection of buildings, as to location and internal arrangement, by which an appropriate environment for each and all is sought to be attained, while at the same time, the opinion gains ground that the domestic or "segregate," as contrasted with the "congregate" or institution idea, should prevail for a large portion, in providing for them economical and substantial buildings, with as much of the house-like and home-like character as in each instance the fact of insanity would permit.

We have here one of the many good evidences that from the first of the organization of the Association, its members have accepted the teachings of Dr. Arnold, that "Nothing is so wrong as the strain to keep things fixed when the whole organization of law and order is one of eternal progress."

Originally the congregate system was naturally adopted in the State lunatic hospitals, as promising to be adequate to all necessities then known. The pressure for admission beyond the original estimates was as unexpected as it was irresistible. The suffering applicants would not and could not be denied. Experience alone can measure the painful perplexities attending the management of a lunatic hospital, and the positive evils resulting from an overcrowded condition. In this connection I cannot forbear recording my admiration of those my justly honored fellow-members of the Association of Superintendents, whose administration under their embarrassments, often with narrow means, has met with such grand success.

Among the thousands of the varied classes of the insane whom the broad charities of the States have so mercifully sheltered, there are many old and hopeless cases, without friends or kindred, the daily care of whose lives has hitherto been often calculated with a rigid economy and scant sympathy, who have a right to claim from their fellow-men a quiet and kindly resting-place on their way to the grave. Some of them are truly good men and women, though the moral accountability of their lives is at an end. " Possible angels in another life " (as some one aptly termed them), they are waiting, sometimes in tumult, sometimes in fear, rarely in peace, that conclusion of life which may be to them the prelude of a better existence. I never looked upon this class without hearty interest. As we scatter flowers over the graves of our friends, and keep their resting-places in decency and order, so should we care for those ever worthy of our love, who were never more in need of our thoughtful and practical sympathy.

JOHN S. BUTLER.

APPENDIX.

(1.) Page 8. In 1845, that enlightened and persevering philanthropist, Lord Ashley, to whom the poor of England are greatly indebted for his able advocacy of their interest, submitted to the House of Commons two bills "for the better care of the insane." On presenting them for consideration he made an able speech replete with valuable information (Editor, *Journal of Insanity*). "It seems unnecessary," he said, "that I should weary the House further, to enforce upon an assembly of educated, humane, and liberal-minded men the necessity for making provisions for those unhappy and destitute beings, who, by a wise though inscrutable dispensation of Providence, have been made subject to this awful calamity, and whose suffering and helpless condition demands that they should receive an unusual share of sympathy from every one of us. But it is remarkable how slow and tedious has been the process whereby we have arrived at the rational and kind mode of treatment which now appears to be recommended to all of us, not only by the dictates of humanity, but also by common-sense. Until the period of the Reformation there is not a single instance of a lunatic asylum being established. Persons of station and wealth were confined in their own houses : and whips, chains, darkness, and solitude were the approved remedies. That practice has indeed descended to our own times ; and Dr. Conolly states that he has formerly witnessed 'humane English physicians daily contemplating helpless insane patients bound hand and foot, and neck and waist, in illness, in pain, and in the agonies of death, without one single touch of compunction, or the slightest approach to feeling of acting either cruelly or unwisely ; they thought it impossible to manage insane people any other way.' " It belonged to the French nation, to the genius of French professors, first to make this mighty advance in the cause of humanity. It was reserved for M. Pinel, the great physician, to achieve this great work.

He undertook what appeared to be the rash enterprise of liberating the dangerous lunatics of the Bicêtre.[1] He made application to the Commune for permission. Couthon offered to accompany him to the great Bedlam of France. They were received by a confused noise. The yells and angry vociferations of three hundred maniacs mixing their sounds with the echo of clanking chains and fetters through the dark and dreary vaults of the prison. Couthon turned away with horror, but permitted the physician to incur the risk of his undertaking.

There were fifty who, he (Pinel) considered, might without danger to the others be unchained ; and he began by releasing twelve, with the sole precaution of having previously prepared the same number of strong waistcoats with long sleeves, which could be tied behind the back if necessary. The first man on whom the experiment was to be tried was an English captain, whose history no one knew, as he had been in chains forty years. He was thought to be one of the most furious among them. The keepers approached him with caution, as he had, in a fit of fury, killed one of them on the spot, with a blow of his manacles. He was chained more rigorously than any of the others. Pinel entered his cell unattended, and calmly said to him : "Captain, I will order your chains to be taken off, and give you liberty to walk in the court, if you will promise me to behave well, and injure no one." "Yes, I promise you," said the maniac, "but you are laughing at me ; you are all too much afraid of me." "I have six men," answered Pinel, "ready to enforce my commands, if necessary. Believe me, then, on my word, I will give you your liberty if you will put on this waistcoat." He submitted to this willingly, without a word. His chains were removed, and the keepers retired, leaving the door of the cell open. He raised himself many times from his seat, but fell again on it, for he had been in a sitting posture so long that he had lost the use of his limbs. In a

---

[1] In "Pinel, a Biographical Study," read before the Academy of Sciences, by Casimer Pinel (his nephew), we find this thrilling relation.

quarter of an hour he succeeded in maintaining his balance, and, with tottering steps, came to the door of his dark cell. His first look was at the sky, and he cried out enthusiastically : "How beautiful ! " During the rest of the day he was constantly in motion, walking up and down the staircases, and uttering short exclamations of delight. In the evening he returned of his own accord to his cell, where a better bed than he had been accustomed to had been prepared for him, and he slept tranquilly. During the two succeeding years, which he spent in the Bicêtre, he had no return of his previous paroxysms, but even rendered himself useful by exercising a kind of authority over the insane patients, whom he ruled in his own fashion.

It was spread abroad that Pinel had released the lunatics from their fetters with bad intentions, and under this pretext a furious mob one day brought him "*à la lanterne.*" Chevingé, an old soldier of the French Guards, rescued him out of their hands, and thus saved his life. This man was one of those lunatics liberated by Pinel, afterwards cured, and ultimately taken into his service.

It is elsewhere recorded that for months after his rescue of Pinel, he procured all the needed supplies of the Bicêtre, under the direction of the doctor, who did not dare to be seen in the street.

In his comprehensive and interesting history of the insane, Dr. Hack Tuke quotes from Dr. Pliny Earle (*American Journal of Insanity*, April, 1856), who says : "It is now very fully demonstrated that the idea of the amelioration of the condition of the insane was original with Pinel and Tuke, and that for some time they were actively pursuing their object, each uninformed of the action of the other. It is no new thing for inventions, discoveries, and innovations upon traditionary practices to originate almost simultaneously in more than one place, showing that they are called for by the times, that they are developments of science and humanity, necessary evolutions of the human mind in its progress towards the unattainable perfect, rather than what may be termed a gigantic or monstrous production of one intellectual genius. Each perceived the wretchedness, the misery, the suffering of the insane around him ; each was moved to compassion ; each resolved to effect a reform in their treatment ; each succeeded. The recognition of services to humanity is due to each. To each we freely accord it."

Dr. Ray, in the same journal, speaks of the founder of the Retreat
as "clear-headed and warm-hearted, one who, true to his faith, con-
ceived the idea that the insane as well as the sane could be best
managed in the spirit of peace and good-will."

(2.) Page 23. In describing the Retreat for the Insane in the
Journal of Insanity for July, 1845, page 67, Dr. Bingham says :

" From the report of the Chaplain we make the following extracts :

" ' To guard against relapse also, it ought never to be forgotten, is
a prominent feature of complete success in the care of the insane.
Self-control, prudence in observing the rules of health, watchfulness
in avoiding those kinds and degrees of excitement which tend to pro-
duce a relapse, calm and equable feelings, just views of life, a con-
scientious performance of duty, regular, useful, and encouraging
employment, cheerful resolution and hope, and, above all, moral and
religious principles,—these should be cultivated with the most assid-
uous care, as they constitute the strongest security against the return
of the distressing malady.   That institution which can best succeed
in furnishing its cured and discharged patients with these elements of
security has attained one of the highest ends, if not the very highest,
to be aimed at in this department of benevolent effort.   To do this
the whole man must be put right, or as near right as can be.   Not
only medical but also moral and religious influences must be brought
to bear upon him, or else he will be healed but in part, and subsequent irregularity, or even deficiency, in the working of one portion
of his system may again derange other portions, and the old, or some
new form, perhaps, of mental aberration be the result.'

" These we consider valuable suggestions.   We have long felt and
taught that ' we had not done a patient all the good we ought by
curing him of one attack, but that we should endeavor so to instruct
him that he may prevent another ; that we believed in man's power
over himself to prevent and control insanity in many instances.'

" But to accomplish this men need instruction, especially those pre-
disposed to insanity, and we know of no one better calculated to aid
in enlightening all such on this important subject than the dis-
tinguished Chaplain to the Retreat, and we indulge the hope that he
will prepare a work on the topics to which he has alluded, embracing
also those errors in education and in the moral training of children

and youth likely to dispose them to violent emotions, and ultimately to insanity. Such a work is much needed, and would, we believe, be of great utility."

These are wise and timely words. This application, ever needed, seems of late less efficiently enforced. The ultimate results of typhus, pneumonia, and especially scarlet-fever and diphtheria, in children, would be far less favorable if the danger from relapses of these insidious diseases were not explained and vigilantly watched. The marked change in the convalescence from insanity increases the inability of the non-experts to measure the danger which may remain. The naturally earnest desire of both friends and patients to hasten the return home, with the frequent sharp economies or the really narrow pecuniary means, combines to defeat the cautions of the physician. It has again and again occurred that patients thus unwisely removed have been brought back within a few weeks or months, in most cases not as insane as on the first admission. Generally their experience had made them wiser. Some frankly confessing their mistake, were ready to give the institution another and fairer trial. In equity two such admissions should count but one. Such lessons, if not as promptly efficient as that taught by the "old barn" (see page 24), often proved as effectual.

The more the aspect of home was given to the Retreat, the greater was the readiness of convalescents to remain, in the good hopes of securing a more thorough and permanent recovery. Sometimes when this had been secured, the patients, finding themselves after leaving us giving way under the oppressions and unavoidable influences of their own home lives, have come back for some possible relief, and have found it to their present content in being admitted not formally, but as visitors for a few days or weeks. This was one of the minor but happy influences of the Retreat.

Insanity is confessedly a most formidable disease. The great question now pressing upon us comes from the masses of incurables overcrowding many of our capacious State hospitals, crippling the means of remedial treatment, as well as that of comfort. How shall this multitude be cared for, and how shall we check their rapid increase? In my belief this mass of human beings, hopeless of cure, rightly demanding kindly care, come far more frequently from the ignorance or neglect of scientifically well-known and preventible causes than from the often accepted one of heredity. Passing by the discussion of the due proportion of these fruitful causes, we must look for our best relief in the future to the higher power of prevention, that best means of averting insanity. To this, better than to the science of remedy, we may confidently look for the instrument to root out its subtle heritage.

(3.) Page 32. In his introduction to Dr. Jacobi's "Hospitals for the insane," 1841, Samuel Tuke says : "The reasons which Dr. Jacobi assigns for restricting establishments for the insane to two hundred patients appear to us very satisfactory—this number being considered as being mainly of the curable class."

Dr. Jacobi strongly condemns the practice of admitting or retaining in such a hospital those incurable lunatics who are affected with such forms of insanity as may render them highly distressing or injurious to those who are yet considered curable.

Mr. Tuke says : "I had many years ago an opportunity of seeing the change from large to small classes, and was confirmed by it in the opinion which I had previously formed on comparing the condition of the large companies of patients in one institution with the smaller divisions in another.

"In the one thirty patients were frequently found in one division, in the other the number in each room rarely if ever exceeded ten. Here I generally found more of the patients engaged in some useful or amusing employment. Every class seemed to form a little family. They observed each other's eccentricities with amusement or pity ; they were interested in some degree in each other's welfare, and contracted attachments or aversions.

"In the large society the difference of character was very striking. I could perceive no attachments, and very little observation of each other. In the midst of society every one seemed in solitude ; conversation or amusement was rarely to be observed—employment never. Each individual was pursuing his own busy cogitations, pacing with restless step from one end of the enclosure to the other, or lolling in slothful apathy upon the benches. It was evident that society could not exist in such a crowd."

Some may be interested in the explanation which Dr. Tuke gives of the origin of the familiar term "Retreat" as applied to a lunatic asylum. He states that "one day the conversation in the family circle turned on the question what name should be given to the proposed institution, when my grandmother, who was very much interested in the establishment, quickly remarked that it should be called a *Retreat*. It was at once seen that feminine instinct had solved the question, and the name was adopted to convey the idea of what such an institution should be, namely,—a place in which the unhappy might find a refuge, a quiet haven in which the shattered bark might find the means of reparation or of safety."

(4.) Page 35. Dr. John Brown, the well-known author of "Rab and His Friends," in the memoirs of his father, says : "From his nervous system and his brain predominating steadily over the rest of his body, he was habitually excessive in his professional work. Thus it was, and thus it ever must be, if the laws of our bodily constitution, laid down by Him who knows our frame, and from whom our substance is not hid, are set at naught, knowingly or not—if knowingly, the act is so much more spiritually bad ; but if not, it is still punished with the same unerring nicety, the same commensurate meting out of the penalty, and paying ' in full tale.' It is a pitiful and sad thing to say, but if my father had not been a prodigal in a true, but very different meaning, if he had not spent his substance, the portion of goods that fell to him, the capital of life given him by God, in what we must believe to have been needless, and therefore preventable excess of effort, we might have had him still with us, shining more and more, and he and they who were with him would have been spared those two years of the valley of the shadow, with its sharp and steady pain, its fallings away of life, its longing for the grave, its sleepless nights and days of weariness and languor, the full expressions of which you will find nowhere but in the Psalms and in Job."

"I have often found that the more the nervous centres are employed in those offices of thought and feeling the most removed from material objects,—the more the nervous energy of the entire nature is concentrated, engrossed, and used up in such offices, so much the more, therefore, are those organs of the body which preside over that organic life common to ourselves and the lowest worm de-

frauded of their necessary nervous food ; and being in the organic
and not in the animal department, and having no voice to tell of their
wants or wrongs till they wake up and annoy their neighbors who
have a voice, that is, who are sensitive to pain, they may have been
long ill before they come into the sphere of consciousness. This is
the true reason, along with want of purity and change of air, want
of exercise, want of shifting the work of the body, why clergymen,
men of letters, and all men of intense mental application are so lia-
ble to be affected with indigestion, constipation, lumbago, and lowness of spirits, black bile, melancholia. The brain may not give
way for long, because for a time the law of exercise strengthens it ;
it is fed high, gets the best of every thing, of blood and nervous
pabulum, and then men have a joy in the victorious work of their
brains, and it has a joy of its own, too, which deludes and misleads.
All this happened to my father. He had no formal disease when he
died, no structural change ; the mechanism was entire, but the motive power was gone, it was expended. The silver cord was not so
much loosed as relaxed. The golden bowl, the pitcher at the
fountain, the wheel at the cistern, were not so much broken as emp-
tied and stayed. The clock had run down before its time, and there
was no one but Him who first wound it up and set it who could
wind it up again ; and this He does not do, because it is His law—an
express injunction from Him—that having measured out to His
creatures each his measure of life, and left him to the freedom of his
own will and the regulations of his reason, He also leaves him to
reap as he sows. . . .

  "Hugh Miller, that remarkable man—who stands alongside of
Burns and Scott, Chalmers and Carlyle, the foremost Scotchmen of
their time—in his life a noble example of what our breed can pro-
duce, of what energy, honesty, intensity, and genius can achieve ;
and in his death (' by suicide '), a terrible example of that revenge
which the body takes upon the soul when brought to bay by its inexorable taskmaster. I need say no more. His story is more tragic
than any tragedy. Would to God it may warn those who come after
to be wise in time, to take the same—I ask no more—care of their
body, which is their servant, their beast of burden, as they would of
their horse. . . . Most men have, and almost every man should

have, a hobby ; it is exercise in a mild way, and does not take him away from home : it diverts him ; and by having a double line of rails, he can manage to keep the permanent way in good condition. A man who has only one object in life, only one line of rails, who exercises only one set of faculties, and these only in one way, will wear himself out much sooner than a man who shunts himself every now and then, and who has trains coming as well as going ; who takes in as well as gives out."

In this connection the following extract from a lecture delivered some years ago before the Young Men's Association of Utica on the physiology of the brain, by Prof. Coventry, seems to be in good and timely order :

"Vague and indistinct notions were long entertained as to the instrumentality of the brain in mental operation. This is well illustrated by the following quotation from Burton's 'Anatomy of Melancholy.' It is from the original writings of Marrilium Fiernus : 'Other men look to their tools : a painter will wash his pencils ; a smith will look well to his anvils, hammers, and forge ; an husbandman will mend his plow iron, and grind his hatchet if it be dull ; a falconer or huntsman will have an especial care of his hawks, hounds, horses, and dogs ; a musician will string and unstring his lute. Only scholars neglect that instrument (their brain and spirit I mean) which they daily use, and by which they range all over the world, and which by much study is consumed. This (he says) dries the brain, extinguishes natural heat, and whilst the spirits are intent on meditation above, in the head, the stomach and liver are left destitute, and thence comes black blood, crudities, and melancholy ; so that sedentary and diligent men, for the most part, spend their fortunes, lose their wits, and often their lives also, and all through immoderate pains and extraordinary studies."

We may smile at his physiology, but so far as he represents the effects of intense application of the mind and sedentary habits, he is undoubtedly correct, and shows the close observer of nature. The following, taken from a recent number of the *London Quarterly Review*, exhibits the modern view of the same subject. Speaking of the education of Lord Dudley, the writer observes : " The irritable susceptibility of the brain was stimulated at the expense of bodily

power and health. His foolish teachers took a pride in his preco-
cious progress which they ought to have kept back. They watered
the forced plant with the blood of life. They encouraged the violation of nature's laws, which are not to be broken in vain. They in-
fringed the condition of conjoint moral and physical existence.
They imprisoned him in a vicious circle, where the overworked brain
injured the stomach, which reacted to the injury of the brain. They
watched the slightest deviation from the rules of logic, and neglected
those of dietetics, to which the former are a farce. They taught him
no exercises but those of Latin ; they gave him a Gradus instead of
a cricket bat, and his mind became too keen for its mortal coil, and
the foundation was laid for ill-health, derangement of stomach,
moral pusillanimity, irresolution, lowness of spirits, and all the pro-
tean miseries of nervous disorders by which his after life was haunted.

" The picture drawn of Lord Dudley's education has its counter-
part in every day's experience. . . . The overwrought and over-
stimulated intellect is literally nourished with the blood of life. The
brain is inordinately excited at the expense of every other part of
the system, and life or permanent ill-health is too often the penalty
paid for this violation of nature's laws."

Some years ago there was published in London, a valuable series
of "Small Books on Great Subjects, Edited by a Few Well-Wishers
to Knowledge."

Among them was one by the Rev. John Barlow, of London, entitled " Man's Power over Himself to Control Insanity."

To the 3rd edition of this valuable essay I am greatly indebted, not
only for the timely suggestions, but for several pertinent quotations
from foreign authorities.

Mr. Barlow quotes[1] with hearty commendation Mr. Gallaudet's
testimony to the beneficial influence of the religious services of the
patients as follows :

" In estimating their value there are many things to be taken into
account in addition to their spiritual benefit to the patient. . . .
Such are the following : the necessary preparations to be made for
attending the religious exercises in a becoming manner, and which

---

[1] Report of the Retreat for 1846.

fill up a portion of time agreeably and profitably ; the regular return of the stated hour for doing this, and the pleasant anticipations connected with it ; the change of scene from the apartments and halls to a commodious, cheerful, and tasteful chapel, there to unite in the worship of God ; the social feelings induced and gratified, the waking up of formerly cherished associations and habits ; the soothing, consoling, and elevating influence of sacred music ; the listening intelligently to the interesting truths of the word of God, and uniting with the heart in rendering him that homage which is his just due, as is, beyond doubt, the case with not a few of the patients ; the successful exercise of self-control, so strikingly and constantly exhibited by those who need to exercise it ; . . .   the habits of punctuality, order, and decorum which they acquire in going to and retiring from the accustomed place of their devotions; . . .   the feeling that in all this they are treated like other folks, and act as other folks do ; and the subsequent satisfaction, a part of our common nature, which many of them experience in the reflection that they have performed an important duty. . . .   All this is frequently and abundantly confirmed by statements on the part of restored patients before leaving the Retreat, who speak with gratitude of the interest they have felt in the religious exercises, and of the comfort and benefit they have derived from them, and from the other means of religious counsel and consolation which they have enjoyed."